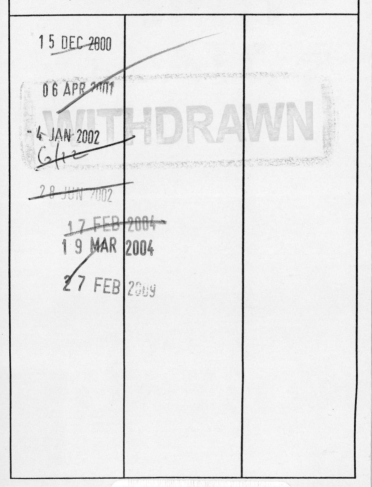

D0412581

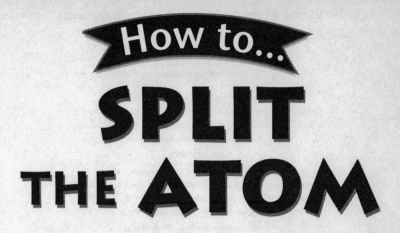

How to...
SPLIT THE ATOM

By HAZEL RICHARDSON

Illustrated by
Scoular Anderson

OXFORD
UNIVERSITY PRESS

For Becky, Cathy and Sam for being great friends

OXFORD
UNIVERSITY PRESS

Great Clarendon Street, Oxford OX2 6DP

Oxford University Press is a department of the University of Oxford.
It furthers the University's objective of excellence in research, scholarship,
and education by publishing worldwide in

Oxford New York

Athens Auckland Bangkok Bogotá Buenos Aires Calcutta
Cape Town Chennai Dar es Salaam Delhi Florence Hong Kong Istanbul
Karachi Kuala Lumpur Madrid Melbourne Mexico City Mumbai
Nairobi Paris São Paulo Singapore Taipei Tokyo Toronto Warsaw

with associated companies in Berlin Ibadan

Oxford is a registered trade mark of Oxford University Press
in the UK and in certain other countries

British Library Cataloguing in Publication Data available

ISBN 0-19-910592-8

1 3 5 7 9 10 8 6 4 2

Printed in the United Kingdom
by Cox & Wyman Ltd, Reading, Berkshire

Contents

HOW TO SPLIT THE ATOM

Atoms are such tiny things that you can't see them, but knowing how to split them up can make you very powerful. The first atom was split early in the 20th century, and it allowed scientists to make nuclear power stations and atomic bombs! Atom splitting is very exciting and very dangerous.

This book tells you everything you need to know about how to become an atom splitter. After reading it, you'll know all about:

What atoms are

How we know that they exist

What atoms are made of

How you can split them

What happens when you split an atom

Where you can find atoms to split

How to protect yourself against deadly radiation

The brilliant scientists who discovered how to split atoms and what they did.

CHAPTER ONE

WHAT ARE ATOMS?

You've never seen an atom, but they're the most important things in your life. Everything is made up of atoms – you, the ground, the air you breathe, the pages in this book, even space.

Atoms are like tiny Lego bricks (but round) that can be joined together in so many different ways that they make up absolutely everything.

How tiny are atoms?...........................

Atoms are so tiny that you could fit a million of them side by side on top of this full stop. A small speck of dust contains one million billion (1,000,000,000,000,000,000) of them! If the

atoms in your body were the size of peas, you would be so big that you would be able to play football with the Sun – wearing heat-proof boots of course!

But, if atoms are that tiny, how do we know that they even exist? It's a good question. People have been asking it for thousands of years. To find out how we know that atoms exist, we have to start over two thousand years ago with a man called Democritus.

7

The Atom Detectives
Part 1: Ancient Greece, around 400 BC

Here we are in Ancient Greece. It's a very good place to live —
if you're not a slave or a woman. The Greeks have so many
slaves that most of the wealthy men do not have to do any
work. They have as much free time as they want to sit around,
chat, drink wine and think about things.

One of these groups of people who don't have much to do call
themselves the atomists. Their leader is Democritus. Today,
they're trying to think about what everything in the universe is
made up of.

People think that everything in the universe is made up of air, water, fire and earth, but I think that this is a load of rubbish!

And that is how atoms got their name.

Ancient atomic ideas ....

The atomists believed that everything was made up of tiny atoms, a different type for every different thing. There were air atoms, cheese atoms – even people atoms. They were partly right, but partly wrong too. You don't see many people atoms around, for example. There were also a few problems with Democritus's idea of cutting things up with a knife:

Atoms are so small that you wouldn't be able to see what you were doing. You would never be able to see where you were supposed to cut.

Because you couldn't see what you were doing, you would never know if you had got down to an atom or not.

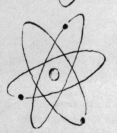

In any case, no knife could cut something so finely that you would be left with just one atom. In fact, by the time you'd cut the cheese into the smallest piece you could manage, there would still be millions of atoms in it!

Because Democritus couldn't prove his idea about atoms, nearly everybody thought that it was just a load of rubbish. But if we jump forward about two thousand years, we can meet two amazing scientists who changed all that. They are Antoine Lavoisier and John Dalton.

The Atom Detectives

Part II: France, a few years before the French Revolution of 1789

Here's Antoine Lavoisier, who shot to fame in 1767 when he came up with a number of bright inventions, including a new idea for street lighting and also Plaster of Paris. Soon after, he received a large inheritance and decided to become a tax collector. This was a very big mistake! The tax collectors were widely hated all through France.

I'm here to collect your taxes, peasant!

Watch out Lavoisier! We peasants have had enough. We will soon rise up in a great and bloody revolution, and have our revenge!

Over the next 25 years, Lavoisier found out how dyes work, how metals rust and how water could be carried on long sea voyages. Then in 1775 he came up with an even more amazing idea.

Everything is made up of elements. These elements can react together to make different things!

Lavoisier wrote a short book about elements and how he thought that they might be made up of smaller things. It would have been a longer book, but something got in the way. It was called the Revolution.

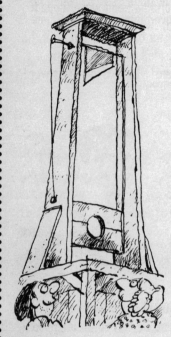

The French peasants finally decided that they'd had enough and started to revolt. Guillotines were built all over Paris and heads began to roll. Royalty and the aristocrats were first, but soon they were chopping off the heads of "enemies of France" — anyone who was unpopular with the revolutionaries. Lavoisier was in trouble. Not only was he a tax collector, but an old enemy, Jean-Paul Marat, was in the Revolutionary government.

Ah! Remember me, Lavoisier?

Oh dear!

Discovering elements...........................

The modern idea of an element was thought up by the English chemist Robert Boyle in the 17th century. He said that an element was a chemical that was made up of only one type of atom and couldn't be broken down into anything else. This was a good idea (and right) but the problem was that nobody was really sure which things were elements and which were not! Scientists knew that gold, iron and lead were elements, but they also thought that mixtures like air were elements. Some of them thought that elements were made up of one type of atom, but they had no way of proving it. Even Isaac Newton (who is famous for discovering gravity) thought that things were made up of atoms. He even worked out how they could bounce and move around! If a great scientist like him could not prove that atoms existed, how would anyone ever be able to do it?

Be a nuclear scientist—
SEE HOW ATOMS MOVE AROUND

Groups of atoms joined together are called molecules. You can see how these joined atom groups move around in this simple experiment using water and food colouring.

WHAT YOU'LL NEED
- ☢ A jam jar or glass
- ☢ Water
- ☢ Food colouring (any colour will do)

WHAT TO DO
1. Pour some water into the jam jar or glass.
2. Add a few drops of the food colouring.
3. Watch what happens.

WHAT HAPPENS?
The molecules in the food colouring move around in the water. As they spread out, the colour spreads through the water, until all the water is coloured. The molecules move all by themselves.

The Atom Detectives
Part III: England, early in the 19th century

Here's John Dalton. He was a very unusual boy, and could go on and on about science for hours. He even started to teach when he was only 12 years old!

In 1803, he came up with the amazing idea that all elements were made up of atoms, and that atoms were solid and hard. Of course, this is exactly the same as what Democritus had said thousands of years before. But Dalton went a bit further.

I am going to work out what all the different atoms weigh!

(Of course, Dalton couldn't weigh single atoms — instead he worked out how much the *same number* of atoms of different elements weighed.)

Dalton's figures for atomic weights were pretty good. But he was completely wrong about some other things. For instance, he thought that only atoms from different elements could join together. We now know that atoms from the same element can join together as well.

Because of Dalton's work, some scientists were persuaded to believe in atoms. Others still thought that it was a load of rubbish — right through to the beginning of the 20th century! Even the famous scientist Ernst Mach (whose name you might have heard of to do with supersonic flight) didn't believe in atoms. He said, "Atoms cannot be seen and are things of the imagination."

Seeing is believing...............................

Today, scientists can actually see atoms, using a microscope so powerful that it makes them look like tiny balls lying in rows. The first pictures of atoms were taken in 1970, and so we know that they definitely exist!

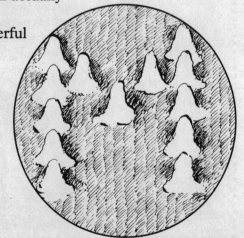

CHAPTER TWO

ATOMS, MOLECULES, ELEMENTS AND COMPOUNDS

So we know that the smallest part of anything is an atom, and that different types of atoms can join together to make different things. But how many different atoms are there?

There are 92 types of atom found naturally in the universe. Each atom has its own name, such as silver or carbon, and a shorter chemical symbol (carbon is C). Jons Berzelius came up with the symbols in the 19th century.

These are too easy — let's put in a hard one!

SHORT NAMES FOR THE ELEMENTS by J. BERZELIUS

CARBON C
OXYGEN O
HYDROGEN H
HELIUM He
SILVER Ag

Scientists have made some new atoms of their own too, because they want to see just how large the biggest atom can be. Some of these created atoms only exist for a hundredth of a second before they disintegrate!

Oops, I blinked!

PING!

Be a nuclear scientist— NAME SOME ATOMS

When scientists create a new atom, they get to give it a name too! All the new atoms have the ending "ium", like "plutonium", for example. Can you work out from the clues what these three atoms are called?

⚛ Atom no. 96   Named after the famous scientist who discovered radium, Marie Curie.

⚛ Atom no. 98   Named after a sunny state on the west coast of the USA.

⚛ Atom no. 99   Named after the most famous physicist of the lot (he said "$E=mc^2$").

Now can you guess which atomic symbol is the right one?

Hydrogen	H or Hg?	Silver	Sl or Ag?
Sodium	Na or S?	Chlorine	Cl or C?
Gold	Ge or Au?	Potassium	K or P?

Answers: curium, californium, einsteinium.
The right symbols are: H, Ag, Na, Cl, Au, and K.

Atoms are friendly!................................

Most atoms don't like hanging around on their own, so they join up with other atoms. Sometimes they join up with other atoms the same as themselves. A material with one kind of atom in it is called an element.

Other times, atoms join up with atoms of a different type.

When atoms of different types join up, they make materials called compounds.

Elementary, my dear Watson! ....

Elements are made up of just one type of atom. For example, gold and silver are elements. Gold is made up of only gold atoms and silver is made up of only silver atoms.

The Element Hall of Fame!

GOLD

This must be one of the most famous elements. It has been used for making jewellery and decorating palaces ever since civilizations began. It's a lovely colour and it never gets rusty or dull. It's rare and expensive, and so people get very greedy about it. One person who got just a bit too greedy for his own good about gold was the legendary King Midas. He was granted one wish by the gods, and immediately asked for everything he touched to be turned into gold. Big mistake! He thought that he would be the richest man in the world, but he died after finding that he couldn't eat or drink anything because it turned to gold! He even turned his daughter into a beautiful golden statue.

OXYGEN

Oxygen is a very important element. How important? Well, you wouldn't be here if it didn't exist! Oxygen is normally a gas (but it can turn into a liquid if it gets very, very cold). Animals — including humans — breathe it in and use it to get energy from food. It would all have been used up ages ago if it wasn't for plants. Plants use carbon dioxide (a gas made up of oxygen atoms and carbon atoms joined together) to make their food. They take the carbon atoms out of it and throw away the oxygen, which we can use. (Another reason to save the rainforests.)

CHLORINE

This is a strange and smelly element. It's a weird green gas and is poisonous to germs. It's used in swimming pools to make sure that you don't catch nasty diseases from any of the other swimmers. It's just a shame that it makes you smell like toilet cleaner afterwards!

sniff-sniff! phew!

MERCURY

This is one of the oddest elements. It is a runny metal. If you've ever dropped a mercury thermometer (not a good idea because the mercury is **very** poisonous) you will have seen it roll around the floor in funny drops.

CARBON

Carbon is another element without which you wouldn't exist. Everything living on Earth is a carbon-based life form. What this means is that all living things on this planet (and maybe some other planets as well!) use carbon as their main building block. But carbon is very strange. Sometimes it is the burnt bit on your toast, the charcoal that will never light when you're trying to start a barbecue, or the slippery grey graphite that you find in the middle of your pencils. At other times, it is the amazingly beautiful and very valuable diamond!

Do you think your cousin would come out on a date?

Cuddly compounds ....

Sometimes, atoms are happiest when they join up
with different atoms. The most important compound
on Earth is water. Water is made up of oxygen atoms
joined together with hydrogen atoms.

Two hydrogen atoms join up with one oxygen atom to
form one water molecule. Scientists write a water
molecule as H_2O. (H is the symbol for hydrogen and
O is the symbol for oxygen.) Now, do you see why
scientists use symbols instead of names for atoms? It
would be so annoying if you had to keep writing "2
atoms of hydrogen and 1 atom of oxygen"!

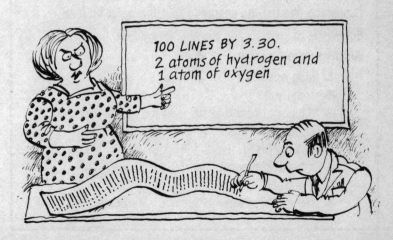

100 LINES BY 3.30.
2 atoms of hydrogen and
1 atom of oxygen

Be a nuclear scientist—
SPLIT A WATER MOLECULE

Molecules are easier to split than atoms. You can see why if you pretend that Lego bricks are atoms. Join two red bricks to a blue brick to pretend they are a water molecule. Now try splitting it up. Easy! You don't need to be very strong to do it. But can you split one of the Lego brick atoms into two bits? Probably not without doing something very dangerous like using a saw — even then it would be difficult.

If we take some water and give it a little bit of energy, we can easily split it up into oxygen and hydrogen. And because oxygen and hydrogen are gases, we can actually see it happening. We get the energy to split the water up from electricity.

WHAT YOU'LL NEED

- ❖ mug or jam jar
- ❖ some salt
- ❖ a piece of card that sits over the top of your mug or jar
- ❖ two pencils
- ❖ a pencil sharpener
- ❖ a battery
- ❖ some copper wire

WHAT TO DO

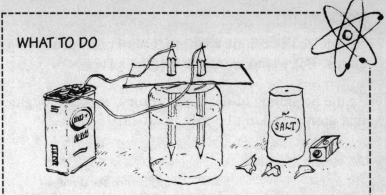

1 Fill the mug or jar with water and add a small amount of salt.
2 Sharpen the two pencils at both ends.
3 Push the pencils through the card and balance it on the jar. (Make sure that the ends of the pencils are in the water.)
4 Connect a battery to the sharp ends of the pencils that are sticking out of the water using the wire.

WHAT HAPPENS?

The beaker explodes

Bubbles of gas collect around the pencils

Nothing

You would probably like the beaker to explode, but it doesn't. Sorry! What does happen is tiny bubbles of gas collect around the pencils in the water: oxygen around one pencil and hydrogen around the other. You have split water molecules into oxygen and hydrogen!

Let's split!...

So, molecules can be split into their constituent atoms. But what can atoms be split into?

At the beginning of the 20th century, people thought that atoms couldn't be split up at all.

Atoms are the smallest building blocks in anything. They're so tiny that there's just no way that you can chop one in half!

Scientists soon had to eat their words, because we now know that atoms *are* made up of even tinier bits...

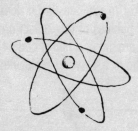

CHAPTER THREE

INSIDE THE ATOM

There are three unbelievably small things that make up atoms: protons, neutrons and electrons.

Here are a proton and a neutron. They're tiny round blobs about the same size as each other. They sit around together in the middle of the atom. This lump is called the nucleus of the atom.

Electrons are 1800 times lighter than protons and neutrons. They zoom around the nucleus, moving so quickly that they're just a blur.

Different types of atom have a different number of protons, neutrons and electrons. For instance, gold atoms have more protons, neutrons and electrons than silver atoms. This is the only difference between gold and silver.

The more protons and neutrons an atom has, the heavier it is.
The lightest atom is hydrogen, which has only one proton and one electron.

What do atoms look like? ....

If we take a closer look at one of the atoms we saw earlier, you will see where the protons, neutrons and electrons fit in.

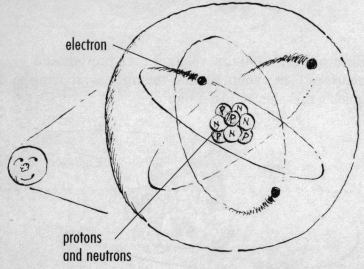

electron

protons and neutrons

The nucleus, in the middle of the atom, holds the other parts together. The electrons zoom around the nucleus on invisible tracks called orbits. If atoms were big enough to see, the speeding electrons would make them look like fuzzy balls.

First impressions

The first scientist to think up this idea of what the atoms looked like was Niels Bohr. Niels was a Danish scientist who won the Nobel Prize for Physics in 1922 for drawing an atom as a picture like this.

Be a nuclear scientist—
MAKE AN ELECTRON PROPELLER

You can get an idea of how fast electrons whizz round by making this electron propeller.

WHAT YOU'LL NEED

- ☢ a piece of card
- ☢ some scissors
- ☢ a pencil
- ☢ a plastic straw
- ☢ some plasticine

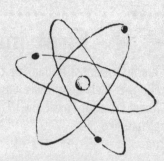

WHAT TO DO

1   Take the pencil and copy this diagram onto your card. (It's easier to do this with tracing paper, if you have some.)

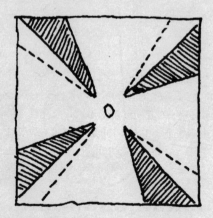

2   Cut along the solid lines.

3   Fold the card over along the dotted lines.

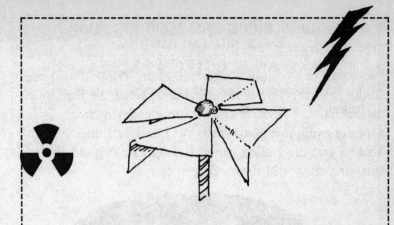

4   Now make a hole in the exact centre of the card using the scissors. (Scientists usually put a piece of plasticine or something else soft under the card to help them do this.)

5   Stick your straw through and hold it in place with a small blob of plasticine.

6   Now, twist the straw.

WHAT HAPPENS?
When you twist the propeller slowly, you can clearly see the separate blades moving. If you spin it fast enough, the lines will all blur together. If you spin it really fast, it will take off, fly round the room, and the top will look like one solid circle! (Make sure that no one's coming into the room when you do this.)

What keeps electrons zooming round the nucleus? ....

Now you're asking a tough question! Think about the Solar System. The planets move around the Sun in a circle because of an invisible force called gravity. Gravity pulls things towards it. The Earth has gravity and so you are pulled down towards it. A good thing too, or you would float off into space!

But if the Earth had too much gravity, it would pull you so hard towards the ground that you would be crushed. Nasty!

An atom works a bit like this, but it isn't gravity that holds the electrons in orbit. It is another force, called electromagnetic attraction.

Sounds complicated? Don't worry, it's very simple. We can see how it works in an experiment using magnets.

Be a nuclear scientist—
FEEL THE FORCE!

WHAT YOU'LL NEED
❖ 2 bar magnets

WHAT TO DO
The ends of bar magnets have different poles. One end is a north pole and the other is a south pole.

Hold the north pole of one magnet close to the south pole of another magnet. What happens?

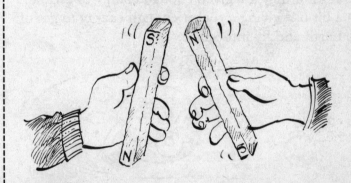

WHAT HAPPENS
The magnets zoom together as the opposite poles attract each other.

Power-packed protons.........................

Remember the clump of neutrons and protons in the nucleus at the centre of the atom? Well, let's forget about the neutrons for now and look at a proton.

Protons have a positive electric charge. This is not the same as the magnetic poles we've just experimented with, but it is similar.

Electrons have a negative electric charge. So, just as the opposite poles of the magnets are attracted to each other, the electrons are attracted to the protons. (But not so strongly that they fly into the nucleus.)

Can electrons ever escape? ....

The force that keeps the electrons in their orbits is not very strong. This means that electrons can get out of the atom if they are given enough energy to escape. It's a bit like giving a rocket enough energy to get off the Earth and fly into space.

Freedom!

When electrons do get away from their atom, they can jump into other atoms. This gives the new atoms a negative electric charge.

I bet you don't know why electricity is called electricity. Do you think that it could have anything to do with electrons? Well, it does! Electricity is just a lot of electrons moving from one place to another. Here's how they were discovered...

An electrifying discovery

People had known about electricity for thousands of years, but nobody knew what caused it. Then, in 1895, a scientist called J.J. Thomson worked out what was going on. J.J. was a very clumsy man. His assistants had to stop him touching any of his equipment, because he always broke it!

J.J. worked out that electricity was a stream of small particles moving from one place to another. He called these particles electrons. He also found out that electrons were 2,000 times smaller than the smallest atom known (hydrogen). This puzzled scientists. If there was something smaller than an atom, could atoms be made up of other particles?

Be a nuclear scientist–
MAKE ELECTRONS MOVE

You can very easily create a kind of electricity called static electricity if you make electrons move from one object to another.

WHAT YOU'LL NEED
- ☢ some pieces of tissue paper
- ☢ a balloon
- ☢ a fluffy duster

WHAT TO DO
1 Tear the pieces of tissue paper into small bits.
2 Blow up the balloon and tie it.
3 Try and pick up the paper with the balloon.
4 Now rub the balloon with the duster quite hard.
  (Not so hard that it pops!)
5 Try and pick up the paper with the balloon again.

WHAT HAPPENS?

When you rub the duster on the balloon, the balloon steals electrons from the duster and gets a small electric charge. This charge on the balloon attracts the pieces of paper.

This works with anything soft and fluffy being rubbed on something hard.

The Ancient Greeks (yes, them again) were the first people to find out about static electricity.

A tricky case to crack..........................

So, you've made electrons move out of an atom! Does this mean that you've split an atom?

Sorry, no. That would be too easy! Electrons are so small that they make up only a tiny bit of the atom. When we split an atom, we have to break open the nucleus.

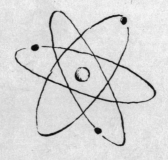

CHAPTER FOUR

RADIOACTIVITY

Electrons are so small and light that most of an atom is made up of the nucleus. To split an atom, the nucleus is the bit that we have to break. When we split an atom, each half of the nucleus takes some electrons with it and makes two completely different new atoms.

Hang on a minute!

If we can only see atoms as balls because the electrons move so fast, how do we know that the nucleus is even there?

And how do we know what it's made of?

We know about the nucleus because of an amazing and very dangerous thing called radioactivity. Let's go back in time again.

The Search For Radioactivity

Part I: Germany, 1789

This scientist is called Martin Klaproth. He's looking very happy because he's just discovered a new element. A new planet called Uranus has also just been discovered in the Solar System.

Hmmm. What shall I call it?

DAILY NEWS
NEW PLANET DISCOVERED

What do you think that Martin will call his new element?

- ⚛ A Klaprothium
- ⚛ B Martinium
- ⚛ C Uranium

Answer: He calls it uranium!

Now let's jump forwards in time, to watch the discovery of a new and exciting ray...

Here's Wilhelm Roentgen, messing around with a cathode ray tube.

A cathode ray tube is a very complicated ray gun type thing. It is usually made of glass and has an electron gun inside it which shoots out rays of electrons. But at this time, nobody knows what the rays are made up of. Some scientists think that they are shooting rays of atoms or molecules.

The end of the cathode ray tube is covered with luminescent paint (the sort of thing you have on clocks with dials that glow in the dark) and the electron beam makes a spot of light on it. If you bend the electron beam around, you can make an image. (This is how the picture on a TV is made.)

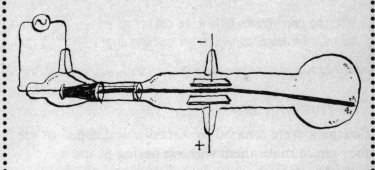

Wilhem is watching what happens as he flashes his new ray gun around, when he notices that the rays make a screen at the other end of the room light up. "I've discovered a new type of ray!" he says. "I'll call them X-rays, because I don't know what they are!"

Wilhelm finds that the rays go straight through light things, but not heavy things. He decides to take a picture of his wife's hand with an X-ray.

Now, don't worry darling, this won't hurt!

Why don't you take a picture of your own hand then?

When he develops the picture, he can see all the bones in her hand, as well as her wedding ring!

The missing link ....

Scientists were amazed by X-rays and decided to see if they could make them without having to use a cathode ray tube.

Now let's jump forward in time again to the following year, to see how trying to make X-rays led to an astounding discovery about uranium.

The Search for Radioactivity
Part III: Paris, 1896

This is Antoine Becquerel, a French scientist who is very interested in X-rays. He's decided to see if he can make them without having to use a cathode ray tube. He's been doing this by leaving lumps of rocks and other stuff on top of photographic film in his garden when it is sunny. Then he sends the film to be developed to see if any rays have been given off. (Rays would make the film go cloudy.)

But today the weather has been pretty rotten. In fact, it's been cloudy and wet for two weeks. (Sounds more like England than France!) Becquerel is disappointed. He's been dying to see what will happen to uranium if he leaves it in the sunlight. Instead, he's had to leave it in a drawer for ages on top of his photographic film.

Anyway today, for some strange reason, he decides to develop the film without putting it in the sunlight. Imagine his amazement when he finds that the uranium has left a mark on the film!

Sacré bleu! This is a new kind of energy ray!

Be a nuclear scientist–
SEE HOW URANIUM CAN LEAVE A MARK ON PHOTOGRAPHIC FILM

Photographs work because the film has a silver compound in it. The silver compound goes dark when it is hit by light, X-rays or radiation. When you get photographs developed you are sent negatives. These are the actual pictures that are taken onto the film. If you hold them up to the light, you can see that the sky is always very dark because of the light hitting the silver compound.

You can see how Becquerel's film looked using light energy instead of radiation.

WHAT YOU'LL NEED
- ☢ a dark room
- ☢ a small coin or other round thing
- ☢ a roll of new photographic film
- ☢ a pair of scissors
- ☢ your parents' permission to chop up a roll of film!

WHAT TO DO

1   Take your new film into the dark room. You must be very careful that no light falls on the end of the film.

2   Using the scissors, cut a small piece of film from the end of the roll.

3   Place the coin in the centre of the film and put it in a place where it won't be knocked.

4   Turn the light on and leave the film for a minute.

5   Turn the light off again and put the piece of film into a light-proof envelope. (One made of card is good.)

6   Get the film developed. You can find a friendly shop that will develop one photograph for you, or you may be really lucky and have a teacher or friend who can develop photographs.

WHAT HAPPENS?

When the photograph is developed, it will show the circle made by the coin. Light didn't hit the film under the coin and so it doesn't change. The negative of the film will look just like the photograph that Becquerel made of his uranium!

Now we have to jump forward in time two years, to when someone discovers what these rays are...

The Search for Radioactivity

Part IV: Paris, 1898

We're still in France. Here's Marie Curie, one of the most famous scientists ever. She and her husband, Pierre Curie, are investigating Becquerel's uranium rays.

The Curies call the rays radioactivity. Then Marie makes an amazing discovery...

Uranium is usually mixed up with other things in rocks. Marie gets her uranium from a rock called pitchblende. But she finds that there is someting else in the pitchblende, giving off even stronger radioactive rays than uranium! After years of work, she and Pierre finally get this new radioactive element out of the rock...

Radioactive risks ....

Just like X-rays, people thought that radium was amazing. It gave off a lovely bright blue glow in the dark. It was used to make luminous clocks and watches. It was put in bath salts to cure aches and pains. It was even used in toothpaste to make your teeth whiter than white. What it didn't tell you on the label was that if you used it for long, you would have no teeth left!

We know today that radioactivity is very dangerous. So dangerous that it can kill you! Marie Curie died of the effects of radioactivity after working with it for years. But until she did, nobody really guessed just how nasty this stuff was.

But how did discovering radioactivity help us find out about the nucleus of an atom?

Well, Marie Curie thought that the radioactivity was coming from the uranium or radium atoms. This caused a real stink amongst the other scientists. How could the atoms give off so much energy?

I think that the atoms are breaking into smaller pieces. These pieces are the radioactive rays!

Don't be daft! Atoms are the smallest pieces of anything! They can't be broken up!

They'd better not be breaking, or I'll have to rewrite my textbook!

The problem was that if atoms could split into smaller bits, scientists would have to change their whole ideas about the atom. Some of them had only just begun to believe in atoms in the first place. Now, they were being asked to change their minds all over again! So Marie Curie's ideas were not popular. But she turned out to be right!

CHAPTER FIVE

THE NUCLEUS

When electrons were first discovered, scientists thought that atoms must be like plum puddings, with the electrons stuck like raisins all over the atom ball.

STRUCTURE OF THE ATOM

Mmm... Maybe it needs some holly on the top.

Now it's time to meet the man who made them change their minds – Ernest Rutherford.

Hi!

Rutherford had been a farmer and mill owner in New Zealand before he went to study in England in 1895. He worked with Joseph Thomson, the man who two years earlier had discovered that electrons existed. Then Rutherford decided to work on radioactivity.

In the radiation given off by uranium, he found two kinds of radioactive ray. He called these alpha rays (after the Greek letter A) and beta rays (after the Greek letter B). Alpha rays were not very powerful and couldn't get through anything more than a piece of paper. But beta rays could get through thin pieces of metal. Rutherford also discovered that the rays were made up of particles, just like the electron ray.

The Search for Radioactivity
Part V: America, 1898

Rutherford has been given a large supply of radium, which is very expensive. He's also met up with Frederick Soddy and the two men have decided to work together. They've just discovered some unbelievable things. They've found that if you let a radioactive element just sit around and throw off its radiation, something very strange and miraculous happens.
The element changes into something else — a new element!

Hey, Soddy, look at this! The thorium has changed into lead!

Great, I could do with some lead to fix my roof!

This makes Rutherford think that radioactivity is bits of an atom. As more bits are thrown out, the atom gets smaller and so it turns into another type of atom. What happens when Rutherford tells other scientists about his idea?

They were absolutely horrified! This couldn't happen!

Still peeved about the reaction from other scientists, Rutherford moves to Manchester in 1907. There, he carries out one of the most important experiments ever done in the entire history of the world. Rutherford blasts a ray of alpha radiation at a piece of gold foil. (Like tin foil to wrap your roast chicken in, but much thinner.)

The results are unbelievable! Most of the alpha radiation goes straight through the foil. This is what Rutherford expects to happen. What he doesn't expect to happen is for some bits of the radiation to bounce back off the foil.

It takes Rutherford a year to work out what caused this amazing result. He tries to imagine what is happening as a particle of alpha radiation goes towards the gold atoms... "Imagine that a burglar is running away after robbing a bank. The police have set up a road block, but there aren't enough policemen to stand side by side across the road. They have large gaps in between them. If the robber can run through the gaps he will. But, if he is running in a straight line and he is going to run into a policeman and be caught, he will stop and run away!"

This is what happened to the alpha radiation. Rutherford worked out that the atoms have large gaps between their centres. Most alpha radiation can run through these gaps. But, if some radiation is on a collision course with the centre of the atom, it is repelled and bounces back.

This means that:

❀ Atoms have a tiny centre surrounded by a very large orbit of electrons.

❀ The centre of the atom has the same electric charge as the alpha radiation.
The same poles on two magnets repel each other — you cannot push them together however hard you try. It is the same with two electric charges.

Rutherford is going to win the Nobel Prize for this amazing discovery. He also finds out what the two kinds of radiation are. Alpha radiation is a chunk breaking off a nucleus. It is made up of two neutrons and two protons. Beta radiation is high-speed electrons.

Do you remember that protons have a positive electric charge? Alpha radiation has a positive charge because of the protons. This means that Rutherford knows that the nucleus in an atom must have a positive electric charge!

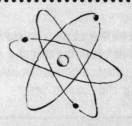

Be a nuclear scientist—
REPEAT RUTHERFORD'S EXPERIMENT

You can see what happened in Rutherford's experiment by using skittles.

WHAT YOU'LL NEED
- some skittles (or jam jars or bottles will do)
- some ping pong balls (or something similar)

WHAT TO DO
1   Line up the skittles with spaces between them. These are the nuclei (plural of nucleus) of the atoms. It's best to do this outside.
2   Roll the balls at the skittles. The balls are the radiation.

WHAT HAPPENS?
Most of the balls will run between the skittle nuclei, but some will bounce back. This is exactly what Rutherford found.

If you picture an atom as big as Wembley Stadium, its nucleus would be about the size of a grain of salt. So imagine how much radiation passes through and how little hits the tiny nucleus and bounces back.

Atoms with a weight problem.................

So, Rutherford had shown that atoms had a tiny nucleus made up of different particles and electrons zooming around outside. He had also shown that atoms could be broken up into other things. In fact, they could sometimes do it all by themselves! But why?

Radioactivity happens when atoms have more particles in the nucleus than they should. This extra weight makes them unstable. What this means is that they're not happy being as heavy as they are and so they often fall apart all by themselves!

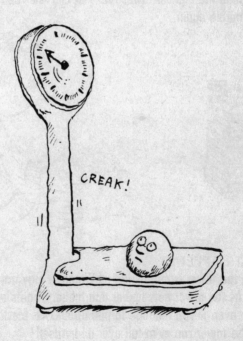

CREAK!

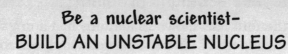

Be a nuclear scientist—
BUILD AN UNSTABLE NUCLEUS

You can see how being too large can make an atom unstable using dominoes.

WHAT YOU'LL NEED
- ☢ some dominoes or wooden building blocks
- ☢ a marble or small ball

WHAT TO DO
1 Pile a few blocks up into a tower.
2 Roll a marble gently at the tower.
3 Add more and more dominoes to the tower and try again.
4 Build the highest tower that you can and then roll the marble again.

WHAT HAPPENS?
When the tower is quite short, the marble may not knock it over. As the tower gets higher and higher, it gets easier to knock over. If you manage to pile up enough blocks, you will find the tower can even fall over all by itself!

Radioactive facts.................................

In some atoms, the neutrons and protons don't stick together properly. This makes the nucleus unstable and the atom radioactive, and it tries to make itself more stable. There are two main ways this can happen:

- The atom can shoot out two protons and two neutrons. This is called alpha radiation.

- The atom can change a neutron into a proton and shoot out a tiny beta particle (the same as an electron). This is called beta radiation.

There are two types of radioactive atom:

1 Extra-large atoms. These are always unstable and radioactive.

Better keep clear of him – he's unstable!

Examples of very large atoms are:

- Radium – The first radioactive element discovered. One kind of radium has 88 protons and 138 neutrons.

- Uranium – The uranium used as a fuel in power stations has 92 protons and 143 neutrons.

- Plutonium – This is one of the deadliest atoms around. One kind is used in nuclear weapons. It has 94 protons and an amazing 145 neutrons!

2 Rare versions of smaller atoms can also be radioactive.

For example, a carbon atom usually has six protons and six neutrons in its nucleus. But there is a much rarer type with six protons and eight neutrons. The extra neutrons make it unstable and radioactive – and not very popular with its friends!

It's smaller atoms like this that become more stable by turning one of their neutrons into a proton and shooting out a beta particle.

If the number of protons in the nucleus of an atom changes, it becomes a different kind of atom. So when radioactive carbon shoots out a beta particle and one of its neutrons turns into a proton, it ends up with seven protons and seven neutrons. It's no longer carbon at all: it has turned into nitrogen!

wheee!

Warning! Really dangerous radiation! .....

Alpha radiation is not very powerful. It can be stopped by a piece of paper. Beta radiation is a bit more powerful. But there is a third kind of radiation that Rutherford didn't know about. This is called gamma radiation. It is not made up of bits of a nucleus – it is waves of energy, like light or X-rays. It travels very fast and can get through very thick lumps of lead or concrete.

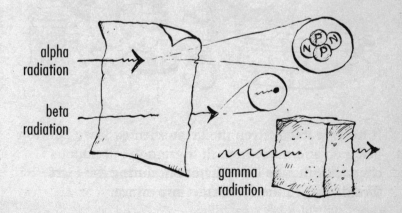

alpha radiation

beta radiation

gamma radiation

Alpha and beta radiation are really only dangerous if they get inside your body. Gamma radiation can go straight through your body and can make you very ill. If you get too much radiation, you soon die.

Turning lead into gold

For thousands of years, people all around the world dreamed of being able to turn cheap metals like lead and iron into gold. These people were called alchemists. They would slave for hours over a fiery furnace, mixing up weird recipes to try to make gold.

It never worked. Even the great scientist Isaac Newton had a go and made himself ill from the poisonous chemicals that he used. Later on, during the First World War, Rutherford had a brainwave.

Hang on!
If radioactive atoms can break down by themselves to make something different, perhaps I can make an atom break down and turn into something else, like gold!

Rutherford did his first experiment on nitrogen. He blasted it with alpha particles and some of the nitrogen atoms split up and threw off hydrogen atoms! This was the first time that an atom had been deliberately split.

But what about making gold? Unfortunately for alchemists everywhere, Rutherford found that making gold out of something else isn't that easy. You have to take an atom that is heavier than gold and knock just the right amount of protons and neutrons out of the nucleus. You *can* do it, but there's a problem.

A lot of people thought that if splitting atoms couldn't be used to make gold, why should they bother doing it? Luckily, Albert Einstein was around.

Can you get energy from splitting an atom? ....

Einstein was right – although not all atoms release energy when they are split, some certainly do. When these kinds of atom split, the two new atoms made are slightly lighter than the original atom. The weight (or mass) that disappears is turned into energy.

Everybody needs energy to run cars, ships and aeroplanes and for electricity. Now there was a chance that just by splitting up atoms into smaller ones, we could have all the energy that we wanted! This was the main reason to try and split the atom. There were only two problems.

1 Sometimes, when you fire particles at an atom like Rutherford did, the nucleus just grabs them and makes itself bigger.

2 The only atoms that give us energy when we split them are some of the large, radioactive ones. And as we know, radioactivity is a horrid thing!

However, in 1932, Otto Hahn and Fritz Straussman managed to split a special type of uranium atom into two bits and showed that energy was released. For this, they won the Nobel Prize and scientists started to work out how to use the energy from atoms to make electricity...

CHAPTER SIX

AND FINALLY...
SPLIT AN ATOM!

Are you ready to split an atom? We know that…

 When you split an atom, you have to break up its nucleus to make two new atoms.

 The reason why we would want to split the atom is to make lots of energy.

 Not every atom can be split, so we have to use radioactive atoms. The one we use most is deadly uranium.

Splitting atoms is very, very dangerous. But if you still want to do it, let's get going.

Step 1 – Find something to split the atoms with ....

Hmm... now which shall I choose?

To split the atoms properly, you need to hit them with something smaller than a nucleus. If you hit them with something bigger than the nucleus, then the atoms will just be pushed around and not split open.

If something smaller than the nucleus hits it, it forces its way in-between the protons and neutrons. In atoms that split, the protons and neutrons don't like this! They split apart.

There are three particles smaller than the nucleus. Which one do you think you should use to split the atom?

⚛   A. Electron?
⚛   B. Neutron?
⚛   C. Proton?

The answer is B. To split atoms you need to fire neutrons at the nucleus using a neutron gun.

Why do I have to use a neutron?

You can't use an electron because they're just too tiny to do the job. Protons are much bigger and they could work if it wasn't for our old friend electromagnetic attraction.

Be a nuclear scientist—
SEE WHY PROTONS ARE NO GOOD FOR SPLITTING ATOMS

Protons have a positive electric charge. Neutrons have no electric charge. In this experiment you can see what would happen if you shot a positively charged proton at a nucleus full of positive charges.

WHAT YOU'LL NEED
- ☢ three magnets
- ☢ some rolled up pieces of paper the same size as the magnets
- ☢ a small elastic band

WHAT TO DO

1 Mark the ends of all the magnets that have the same pole. You can do this by seeing whether they attract (pull towards) or repel (push away) each other. If two poles repel, they have the same magnetic charge.

2 Put two of the magnets and two of the rolled up pieces of paper together as in the picture and fasten them with an elastic band. Make sure that your marked poles are both at the same end. This is your nucleus.

3 Hold the nucleus in one hand with the marked poles facing towards you and try to push the marked end of the other magnet into the nucleus.

WHAT HAPPENS?

The magnet cannot be forced into the nucleus because it has the same charge as the magnets facing it.

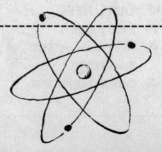

No charge

A proton would be no good for splitting an atom because it has the same charge as the protons in the nucleus, so it would be repelled. A neutron doesn't have an electric charge and so it doesn't have this problem.

When a neutron is fired at a uranium nucleus using a neutron source, the nucleus captures it. It stretches out to try and fit the neutron in. The nucleus cannot handle the extra neutron and gets thinner around the middle. Then it splits open into two parts. Each half flies out at great speed.

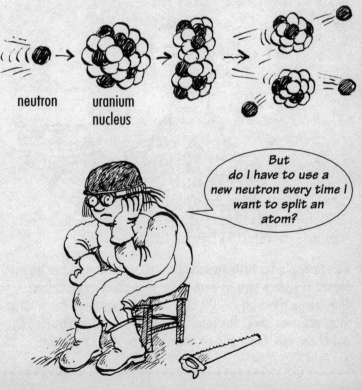

neutron

uranium nucleus

But do I have to use a new neutron every time I want to split an atom?

That would make splitting a lot of atoms very difficult. But luckily, something amazing called a chain reaction means that you don't have to. One neutron can be used to split millions of atoms! To find out what a chain reaction is, let's go back in time again…

Starting a chain
America, 1942

The Second World War is on, and American scientists are very interested in the power released by splitting atoms…

This is the Italian scientist Enrico Fermi.

Hi!

He's just split his first uranium atom up. It turns into two atoms almost the same size as each other — caesium and rubidium. The energy given off is 200,000 volts of electricity! But what is most amazing about the reaction is that a few neutrons also fly out of the split atom and go zooming around on their own.

Enrico decides that he can slow down the neutrons by making them bounce off other atoms that won't be split up by them. He finds a disused squash court at the University of Chicago and builds the world's first nuclear reactor (he called it an "atomic pile") by putting lumps of uranium into a stack of graphite (a form of carbon).

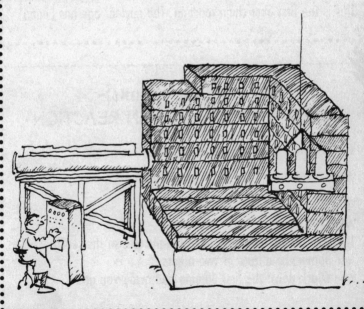

Fermi finds that he doesn't need to bombard the pile with neutrons to get the chain reaction going. There are enough stray neutrons in the air to start the reaction off. The first uranium atom splits and two neutrons fly out. The carbon atoms slow down the neutrons, and they hit two more uranium atoms. These two uranium atoms send out four more neutrons, which hit four more uranium atoms, and so on.

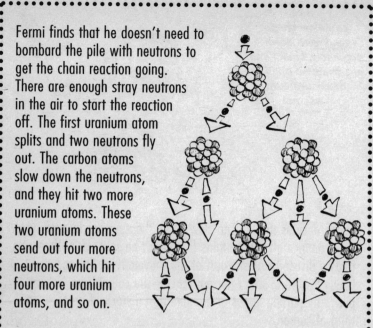

This is the first ever chain reaction. The nuclear age has begun!

Be a nuclear scientist—
BUILD YOUR OWN CHAIN REACTION

You can build your own chain reaction. Here's how.

WHAT YOU'LL NEED
☢ lots of dominoes

WHAT TO DO
1   Line up the dominoes as in the picture, so that each domino is in front of two others.
2   Knock down the first domino and away you go!

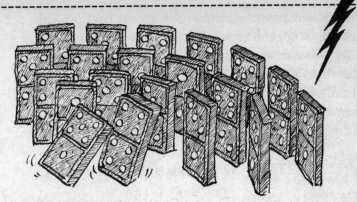

WHAT HAPPENS?
After you push the first domino over, each domino that falls knocks over another two dominoes. This is very like what happens when you split atoms in a nuclear reaction.

Chain reactions don't happen with every type of atom. To make a chain reaction to build a nuclear power station, you need to find the right kind of atoms!

Step 2 – Find your atoms.....................

In nuclear power stations, the atoms that are split are usually a special type of uranium atom. Remember that you can get isotopes of atoms, with different numbers of neutrons in the nucleus? Well, most uranium is called uranium-238. The number after the name tells you the number of protons plus the number of neutrons in the nucleus.

Uranium-238 has 92 protons and 146 neutrons.

$$238 = 92 + 146$$

Uranium-238 is radioactive, but it does not split up when you bombard it with neutrons. Instead, it just absorbs them.

The special type of uranium atom that *can* be split is called uranium-235. It has 92 protons and 143 neutrons. Unfortunately, this type of uranium is much rarer than uranium-238. Because it's so dangerous, it's not sold in chemists' shops or in the supermarket.

Fermi had to slow down the neutrons to get a chain reaction because fast neutrons are easily absorbed by uranium-238. Slow ones can dodge the uranium-238 and go on to split uranium-235.

And you can't go for a walk and just hope to find a lump of it lying around! Uranium is never found on Earth as pure uranium. It is always joined up with other atoms in rocks. One rock it is found in is called pitchblende. Another is called carnotite. Pitchblende is mostly found in Canada, Zaire, and the United States. In America, there are several carnotite mines in Colorado, Utah, New Mexico, Arizona, and Wyoming. These mines are guarded and the amount of rock taken is measured carefully to make sure none is stolen. So it is very difficult to find any uranium to split unless you are working as an atom-splitting scientist or running a nuclear power station.

But if you ever do manage to find some pitchblende, you have to get the pure uranium out of it before you can split it up. The first thing to do is to break the pitchblende rocks into smaller pieces. Then treat the pieces with with nitric and sulphuric acids to dissolve the uranium inside them.

Acids are very strong chemicals and very dangerous when they are not diluted. Some acids are strong enough to dissolve a human body – something that some people have used to try and get away with murder!

To get the uranium out of the acid, you have to use another strong and dangerous chemical called sodium hydroxide. This is an alkali.

When acids and alkalis are put together, something very interesting happens. The acid and alkali cancel each other out, so that the acid isn't acid anymore. When the acid with the uranium dissolved in it is added to sodium hydroxide, the acid and alkali react together. The uranium comes out of the acid as yellow-coloured crystals.

There's just one more thing that you have to do first. The uranium you've got from your acid and alkali reaction is very radioactive, but not all of it can be split up. Most of your uranium will be the uranium-238, which you can't use to build your nuclear reactor. Only a few of the atoms in your lump of uranium are the sort you can split.

Step 3 – Get the right uranium atoms....

The only difference between uranium-238 and uranium-235 is the number of neutrons that they have. The three extra neutrons make the uranium-238 atom a tiny bit heavier than the 235 atom. Because of this, we can separate them!

The easiest way to do this is to make the uranium into a gas. The uranium that you have can be joined to fluorine to make uranium fluoride. You've probably heard of fluoride because it is put into toothpaste and water supplies to help make your teeth strong. Uranium fluoride is a gas.

The uranium fluoride gas is pumped against barriers that have millions of tiny holes in them. The uranium-235 atoms are slightly lighter than the uranium-238 atoms. This means that molecules of the gas made with them go through the holes faster than the molecules made with uranium-238 atoms.

After going through thousands of barriers, the uranium at the other end has much more uranium-235 in it than it had before.

Be a nuclear scientist—
SEPARATE HEAVY AND LIGHT THINGS USING HOLES

This experiment will not only help you see how the different types of uranium are sorted, it will also help you get into your parents' good books if you spill the contents of the store cupboard over the floor.

WHAT YOU'LL NEED
- ☢ a deep bowl
- ☢ some flour
- ☢ some sugar
- ☢ weighing scales
- ☢ a pin
- ☢ a pair of scissors
- ☢ a large elastic band
- ☢ some greaseproof paper

WHAT TO DO
1 Cut a piece of greaseproof paper that is large enough to sit over the dish with some hanging over.
2 Fasten the greaseproof paper around the dish with the elastic band, so that the paper sits straight on top of it.

3　Using the pin, carefully make lots of tiny holes in the paper.
4　Weigh a tablespoon of flour.
5　Weigh a tablespoon of sugar.
6　Mix the flour and sugar together very well.
7　Put a spoonful of the mixture on top of the paper and rub it around gently with your finger.
8　Take the paper off the dish carefully and see what is inside.

WHAT HAPPENS?

The flour is lighter than the sugar and slightly smaller as well. It goes through the holes quicker than the flour. If you did this lots of times, you would be able to separate most of the flour out of the sugar. (But don't try it — you'd get very bored!)

Is that it? NOW can I split an atom?

Yes, when you've got enough uranium atoms, you can use them in a nuclear reaction.

Step 4 – Split an atom! ....

Here's the shopping list you'll need in order to build a nuclear reactor to split your atoms in.

MY REACTOR SHOPPING LIST

Nuclear fuel – uranium with lots of uranium-235 atoms

A neutron gun – to fire the neutrons at the uranium

A moderator – to slow the neutrons down

Control rods – to stop the chain reaction if things go wrong

A water supply (and another emergency source) – to cool the reactor down

A huge building – to house it all in

And here's how to put the bits together to make an atom-splitting reactor.

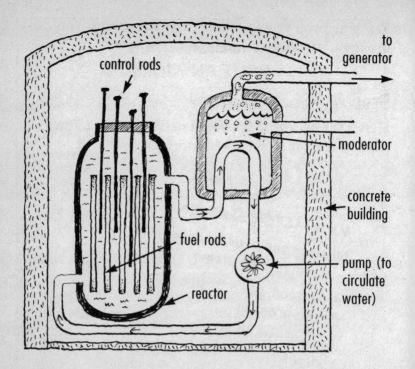

(This is a water-cooled reactor.)

OK - get splitting!

... By the way, just in case you can't lay your hands on any of these vital things, you can see what happens when you split an atom by doing this experiment.

Be a nuclear scientist–
SPLIT AN ATOM!

WHAT YOU'LL NEED

- ☢ five large pingpong balls
- ☢ a thin elastic band
- ☢ some double sided sticky tape
- ☢ thick clothing and safety goggles
- ☢ a hammer, mallet, or something similar
- ☢ someone to help you put your atom together

WHAT TO DO

1. Stick two balls together on one side using the tape. Do the same for another two balls.

2. Hold the four balls together so that they make a square and ask someone to put the elastic band around the balls, as in the picture. You need a thin elastic band that you can only just fit around the balls. This is your atom.

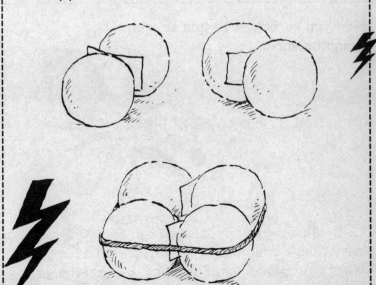

3   Now, put your goggles on
    and take your atom
    outside.
4   Put the other ball on top of
    the atom so that it sits on
    the space between the
    neutrons and protons. This
    ball is your neutron.
5   Carefully use the hammer
    to push the neutron into
    the atom!

WHAT HAPPENS?
The bonds holding the neutrons and protons together break
open as you force the neutron in. The atom splits!

Now you know how to split an atom.
Congratulations!

CHAPTER SEVEN

NUCLEAR POWER – THE PROS AND CONS

Because a nuclear reaction is so fast and powerful, things can sometimes go dreadfully, horribly wrong! Even when you've built the safest power station you can think of, things can still go wrong. Disasters can happen because the people running the power station get tired and don't pay attention to what is going on. If you don't pay attention to what your teacher tells you in class, it's not going to cause a disaster, but if you don't spot that the reaction in your power station is going out of control, you can have a nuclear meltdown!

Help! She's heading for meltdown!

Three Mile Island, Pennsylvania

28TH MARCH 1979

Everything seems to be going well at the Three Mile Island Power plant but, unknown to the workers checking the reactor, a light isn't working properly. A valve in a pipe taking cooling water to the reactor core has been left shut, but the light says that it is open. The workers have no idea that anything is wrong.

The reactor gets hotter and hotter. Eventually, it gets so hot that it melts! Luckily, nearly all the radioactivity is kept inside the reactor building, so not much damage is done. But the reactor is wrecked!

Chernobyl, Ukraine

25TH APRIL 1986

The warnings given after the disaster at Three Mile Island have not been heeded in the Ukrainian power station at Chernobyl. Chernobyl is badly designed and not as safe as power stations in other industrialised countries.

It's early in the morning and yet again, something is going wrong with the cooling system in a nuclear reactor. The workers desperately try to use the water level and the control rods to slow down the reaction and cool the reactor, but within minutes the chain reaction is out of control! It is too late for an emergency shutdown. There is an enormous explosion that rips the reactor open and tears the roof off the building.

Thousands of people have died and are still dying from the effects of the radiation.

More problems with splitting atoms.......

Another problem is what to do with the waste that's left over from the uranium that is split up. You can't just put it into a dustbin or throw it on the tip, because the radiation is so deadly.

There are three levels of waste to deal with:

 Low level waste – paper towels, cardboard and protective clothing used by people working at the power station

 Intermediate level waste – used bits from the reactor

High level waste – this is the used-up uranium fuel.

Solid low level waste can be stored in concrete trenches. Low level liquid waste is simply drained into the sea. This must make the fish very happy!

Intermediate level waste is packed in concrete in large metal drums designed not to be broken open.

The most dangerous waste, high level waste, is a liquid. In Britain, this is turned into large glass blocks and stored. As time goes by and the radioactive elements in the waste break down more and more, they become less radioactive. But some of the radioactive waste may still be deadly in a million years' time! The only thing we can possibly do with it is bury it very deep in the ground and hope that our distant descendants don't stumble on it.

Another idea is to send it into space. This might be the answer – if there are no little green men out there!

The only good thing about nuclear waste is that there isn't a lot of it. This is because even after you've sorted out the lighter uranium atoms that have split up, the uranium in the power station is still mainly the heavier uranium that can't be split. But this heavy uranium is used in another way...

Making plutonium ....

When neutrons are fired at uranium-238 atoms
(which don't split), the neutron just nudges itself into
the nucleus and makes itself comfortable. But not for
long. This extra neutron makes the uranium even
more unstable and it breaks down by itself into
neptunium. Neptunium is also very unstable and very
quickly breaks down into plutonium – a very deadly
radioactive metal in more than one way. Plutonium is
used to make nuclear weapons.

Atomic bombs ....

Once scientists realised the enormous amount of
power that they could get from splitting atoms in a
chain reaction, it wasn't long before this wonderful
new discovery was used to make a new and very
terrifying weapon – an atomic bomb.

The reactions in an atom bomb are the same as in a nuclear reactor. The only difference is that the reaction in the bomb is not controlled with a moderator. The atoms are split into two and neutrons keep the chain reaction going. But in the bomb, the chain reaction only lasts for a short time. The temperature in a bomb gets up to tens of millions of degrees C! Boiling water is only 100°C. The explosion is so powerful that superheated air rushes along for miles, destroying entire buildings. The light from the explosion is so bright that it will blind you if you look at it.

It was only three years after the first chain reaction took place, when the first atomic bomb was tested on July 16th 1945 in New Mexico. In this bomb, the chain reaction only lasted for about a millionth of a second and released enormous amounts of heat energy. If you survive an atom bomb blast, the radiation poisoning from plutonium is worse than uranium.

Nuclear war

Atom bombs have only been used in a war twice. At the end of the Second World War, the Americans decided to try out two atom bombs by dropping them on Japanese cities. This was done to try and speed up the end of the war and save thousands of soldiers' lives.

HIROSHIMA, 6TH AUGUST 1945

The first atom bomb dropped was a bomb made of uranium. It killed and injured over 140,000 people. The bomb was called Little Boy and had the same power as 12,500 tons of TNT.

NAGASAKI, 9TH AUGUST 1945

Only three days after the first bomb was dropped, a plutonium bomb was dropped on Nagasaki. It fell nearly two miles off target and so, even though it had the power of 20,000 tons of TNT, fewer people were killed than at Hiroshima.

Nuclear fusion ....

Atom bombs were powerful, but now we have developed even more deadly nuclear weapons. These work by making light atoms like hydrogen join together to make one bigger atom. Joining light atoms together is called nuclear fusion. Nuclear fusion is what powers the Sun and all the stars, so it's not surprising that nuclear fusion bombs release as much energy as millions of tons of TNT!

So now we know how to split an atom and how to use the energy from splitting it. But it seems too dangerous to do it. Can atom splitting be useful too?

Of course! It has helped us to:

 understand what the universe is made of and how things work.

 build nuclear power stations for electricity, which may be all we have to depend on for a lot of our power when coal, gas and oil run out.

 understand how to join small atoms together in nuclear fusion.

At the moment, we can only use nuclear fusion in bombs, but scientists are trying to work out how to build a fusion power station. Nuclear fusion may well save the world when natural fuels run out. There would be no dangerous waste to get rid of, and it would give us all the power we would ever need.